NATIONAL GEOGRAPHIC KiDS

美国国家地理 趣味双语小百科

原版引进 汉英对照

狼和可爱的朋友们

The Wolf and Lovely Friends

[美] 美国国家地理杂志儿童版 / 编写　付灿 / 译

四川少年儿童出版社

HA! HA!
HA! HA!

加拉帕戈斯象龟的寿命超过100年，体重超过800磅（约362千克）！

The Galápagos tortoise can live to be over 100 years old and weigh more than 800 pounds!

HA! HA! HA!
当当当！
KNOCK, KNOCK.
Who's there? Spine.
Spine who? Quit spine on me!
谁在敲门？斯派恩。
哪个斯派恩？别再暗中监视我了！
*Spine（斯派恩）谐音spying（监视）。

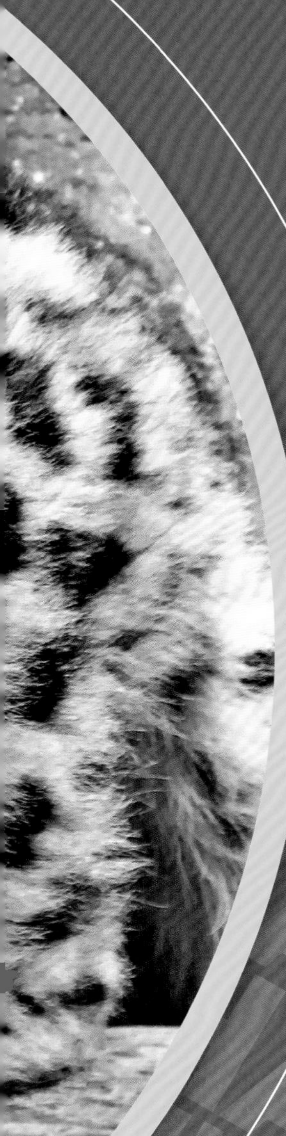

ENCYCLOPEDIA OF ANIMALS 动物小百科

雪豹宽大的脚掌能帮助它们在雪地上行走。

Snow leopards have wide paws to help them walk on top of the snow.

TONGUE TWISTER 英语绕口令

Say this fast three times
快速念3遍

Seventy-seven silly superstitions.

HA! HA! HA! HA! HA! HA!

ENCYCLOPEDIA OF ANIMALS 动物小百科

海豹能在水下停留长达30分钟。

Seals can stay underwater for up to 30 minutes.

Q Why can’t you call the zoo on the phone?

为什么动物园的电话总是打不通？

A Because the lion is always busy.

因为狮子总是很忙。

*lion（狮子）谐音line（电话线路），即“电话线路总是很忙”。

HA!HA!HA!HA!

HA!HA!HA!HA!HA!HA!HA!HA!HA!HA!HA!
当当当！
KNOCK,KNOCK.
Who's there? Anakin.
Anakin who? Anakin you can do, I can do better.
谁在敲门？阿纳金。
哪个阿纳金？任何你能做的事，我都能做得更好。
*Anakin（阿纳金）谐音anything（任何事情）。

这条“美女蛇”属于黑眉锦蛇（中国）台湾亚种，因其身上绚丽的色彩和美丽的花纹而得名。

This looker is a Taiwan beauty snake. They are named for their colorful and beautifully patterned bodies.

随着年龄的增长，大眼树蛙身体的颜色会逐渐由绿色变成褐色。

The big-eyed tree frog turns from green to brown as it grows older.

棉顶狨猴是世界珍稀濒危灵长类动物之一。

Cotton-top tamarins are one of the most endangered primates in the world.

当当当！
KNOCK,KNOCK.
Who's there? Anita. Anita who? Anita borrow your phone.
谁在敲门？安妮塔。哪个安妮塔？我需要借用你的手机。
*Anita（安妮塔）谐音I need to（我需要）。
HA!HA!HA!

HA! HA! HA!
当当当！
KNOCK, KNOCK.
Who's there? Whale.
Whale who? Whale you marry me?
谁在敲门？惠尔。
哪个惠尔？你愿意嫁给我吗？
*Whale（惠尔）谐音will（愿意）。

ENCYCLOPEDIA OF ANIMALS 动物小百科

负鼠的尾巴和身体一样长。

A mouse opossum’s tail is as long as its body.

Q What did the lamp say to the man?

灯对人说了什么？

A 什么也没说，灯不会说话。

Nothing, lamps can’t talk.

HA! HA! HA! HA! HA! HA! HA!
当当当！
KNOCK, KNOCK.
Who's there? Cannoli. *Cannoli who?* I cannoli imagine you aren't ready to go yet?
谁在敲门？卡诺里。
哪个卡诺里？你现在还没准备好出发？我无法想象。
*Cannoli（卡诺里）谐音can not（不能）。

美洲水貂在受到威胁或惊吓时会释放出一种难闻的气味。

The American mink gives off an unpleasant odor if it is threatened or scared.

HA! HA! HA! HA!
当当当！
KNOCK, KNOCK.
Who's there? Yam.
Yam who? I yam so happy to see you!
谁在敲门？扬。
哪个扬？见到你我真高兴！
*Yam（扬）谐音am（是）。

ENCYCLOPEDIA OF ANIMALS 动物小百科

大象的鼻子不仅很有力，能举起重达450磅（约204千克）重的东西，还能灵巧地摘下一片草叶。

An elephant's trunk is strong enough to lift up to 450 pounds and agile enough to pluck a blade of grass.

Q

What runs around a backyard?

什么东西绕着院子跑？

A fence.
栅栏。

当当当！

KNOCK, KNOCK.

Who's there? Odor.
Odor who? I'd like to odor a large pepperoni pizza.

谁在敲门？ 欧多尔。
哪个欧多尔？ 我想订一份大号意大利辣香肠比萨。

*Odor（欧多尔）谐音order（订购）。

南方地犀鸟依赖父母生活的时间长达两年，比其他任何鸟都长。

The southern groundhornbill can depend on its parents for up to two years, longer than any other bird.

当当当！
KNOCK,KNOCK.
Who's there? Weeder. Weeder who? Take me to your weeder.
谁在敲门？威德尔。哪个威德尔？带我去见你的领导。
*Weeder（威德尔）谐音leader（领导）。

ENCYCLOPEDIA OF ANIMALS 动物小百科

狼是体型最大的犬科动物。灰狼一顿能吃掉20磅（约9千克）肉。

Wolves are the largest member of the dog family. The gray wolf can eat up to 20 pounds of meat in one meal.

小马岛猬看起来像一只刺猬，但它实际上并不属于刺猬家族。

The lesser hedgehog tenrec may look like a hedgehog, but it is actually not part of the hedgehog family.

当当当！

KNOCK, KNOCK.

Who's there? Rome.
Rome who? Rome is where the heart is.

谁在敲门？ 罗梅。
哪个罗梅？ 心在哪里，家就在哪里。

*Rome（罗梅）谐音home（家）。

HA! HA! HA! HA! HA!
当当当！
KNOCK,KNOCK.
Who's there? Sven.
Sven who? Sven are we leaving?
谁在敲门？斯文。
哪个斯文？我们什么时候走？
*Sven（斯文）谐音when（什么时候）。

ENCYCLOPEDIA OF ANIMALS 动物小百科

蛇鳗喜欢把身子埋在沙里，伸出头寻找食物。

The snake eel likes to bury itself in the sand and pop its head out in search of food.

TONGUE TWISTER 英语绕口令

Say this fast three times
快速念3遍

My newfangled
bangles get
jangled and tangled.

北极狐有惊人的听力。它们能听见猎物在雪地下挖洞的声音。

Arctic foxes have incredible hearing. They can hear prey tunneling underneath the snow.

HA!HA!
HA!HA!HA!
当当当！
KNOCK,KNOCK.
Who's there? Blur.
Blur who? Blur, it's cold out here!
谁在敲门？布勒。
哪个布勒？呵，外面太冷了！
*Blur（布勒）谐音brrr（呵，表示感觉寒冷）。

当当当！
KNOCK,KNOCK.
Who's there? Bison.
Bison who? Wanna bison Girl Scout cookies?
谁在敲门？比森。
哪个比森？想买点儿女童子军饼干吗？
*Bison（比森）谐音buy some（买点儿）。

ENCYCLOPEDIA OF ANIMALS 动物小百科

环尾长鼻浣熊会发出呜呜声和咔嗒声来提醒同伴注意捕食者。

The ring-tailed coati makes woofing and clicking noises to alert others to predators.

TONGUE TWISTER 英语绕口令

Say this fast three times
快速念3遍

I ate at eight but snacked at seven and eleven.

HA!HA!HA!HA!HA!HA!HA!HA!HA!HA!HA!HA!

ENCYCLOPEDIA OF ANIMALS 动物小百科

大蓝鹭是北美最大的鹭。它的翼展为36到54英寸(约91到137厘米)。

The great blue heron is the largest North American heron. It has a wingspan of 36 to 54 inches.

TONGUE TWISTER 英语绕口令

Say this fast three times
快速念3遍

Vera varnished while Pam polished.

HA! HA! HA! HA! HA! HA! HA! HA! HA! HA! HA! HA! HA! HA!

HA!HA!HA!HA!HA!HA!HA!HA!HA!HA!HA!HA!HA!HA!HA!HA!HA!
当当当！
KNOCK,KNOCK.
Who's there? Raisin.
Raisin who? Is there a raisin you aren't opening this door?
谁在敲门？雷森。
哪个雷森？你不开门有什么原因吗？
*Raisin（雷森）谐音reason（原因）。

HA! HA! HA! HA! HA! HA! HA!
当当当！
KNOCK,KNOCK.
Who's there? Ron.
Ron who? I tried to call
but got the ron number!
谁在敲门？罗恩。
哪个罗恩？我想打电话，
但是号码错了！
*Ron（罗恩）
谐音wrong（错误的）。

ENCYCLOPEDIA OF ANIMALS 动物小百科

威德尔海豹能潜入深达2400英尺（约731米）的水下，并且憋气长达80分钟。

Weddell seals can dive to depths up to 2,400 feet and hold their breath for up to 80 minutes.

TONGUE TWISTER 英语绕口令

Say this fast three times
快速念3遍

Randy rode a raft down a rapid rushing river.

HA! HA! HA! HA! HA! HA! HA! HA! HA! HA!

HA! HA! HA! HA!
当当当！
KNOCK, KNOCK.
Who's there? Sid.
Sid who? Sid down and have a cup of tea.
谁在敲门？锡德。
哪个锡德？坐下喝杯茶。
*Sid（锡德）谐音sit（坐）。

ENCYCLOPEDIA OF ANIMALS 动物小百科

鲸必须醒着才能呼吸。这意味着它们永远无法完全入睡，否则就会淹死。所以，鲸休息时只有半个大脑在睡觉。

Whales must be awake to breathe. This means they can never fall completely asleep or they will drown. Instead, they rest one half of their brain at a time.

TONGUE TWISTER 英语绕口令

Say this fast three times
快速念3遍

Chipper chappy flipper flappy.

当当当！
KNOCK,KNOCK.
Who's there? Alvin. Alvin who? Alvin this race!
谁在敲门？阿尔文。哪个阿尔文？我将会赢得这场比赛！
*Alvin（阿尔文）谐音I'll win（我将会赢）。
HA! HA! HA! HA!

猩猩幼崽会和妈妈一起
生活11到12年。

Young orangutans will
live with their mothers
for 11 to 12 years.

How do you catch a squirrel?

你怎样才能抓住一只松鼠？

Climb up a tree and act nuts.
爬上树，装扮成坚果。

松鼠从100英尺（约30米）高的地方跳下来也不会受伤！它们把尾巴当降落伞。

Squirrels can fall up to 100 feet without getting hurt! They use their tails as a parachute.

HA! HA! HA!
当当当！
KNOCK,KNOCK.
Who's there? Sherry.
Sherry who? Sherry your lunch? I'm starving!
谁在敲门？谢丽。
哪个谢丽？能分享你的午餐吗？我要饿死了！
*Sherry（谢丽）谐音sharing（分享）。

ENCYCLOPEDIA OF ANIMALS 动物小百科

小鹿可以在出生后半个小时内迈出第一步。

Baby deer can take their first steps within a half hour of their birth.

TONGUE TWISTER 英语绕口令

Say this fast three times
快速念3遍

Mortal thwarter.

千万别惊吓灰熊妈妈！灰熊在保护幼崽时以凶猛著称。

Don't surprise a momma grizzly bear! Grizzlies are known to be ferocious when protecting their young.

每只老虎身上的条纹图案都是独一无二的。

Every tiger’s stripe pattern is unique.

大熊猫的食物几乎完全由竹子构成。

A panda’ s diet is made up almost entirely of bamboo.

当当当！
KNOCK,KNOCK.
Who's there? Budda.
Budda who? Budda this toast for me, please.
谁在敲门？布达。
哪个布达？请帮我预订这个吐司。
*Budda（布达）谐音book（预订）。
HA!HA!HA!

ENCYCLOPEDIA OF ANIMALS 动物小百科

鹈鹕的嘴很长，下嘴壳与皮肤相连形成的喉囊可以舀鱼。它们一次能吃很多鱼。

Pelicans use their long bills with expandable attached pouches to scoop up fish. They can eat many fish at a time.

TONGUE TWISTER 英语绕口令

Say this fast three times
快速念3遍

Four furious fiends fought for freedom.

当当当！
KNOCK, KNOCK.
Who's there? Stan.
Stan who? Stan back and I'll try the door again!
*谁在敲门？*斯坦。
*哪个斯坦？*退后，我重新试着开一下门！
* Stan（斯坦），stan back谐音stand back（退后）。

当当当!
KNOCK,KNOCK.
Who's there? Athena. Athena who? Athena shooting star!
谁在敲门? 雅典娜。哪个雅典娜? 我看见一颗流星!
*Athena(雅典娜)谐音I see a(我看见一颗)。
HA! HA! HA! HA! HA! HA!

大壁虎进攻时非常凶猛。它们以昆虫为食。

Tokay geckos have a vicious bite. they eat insects.

HA! HA! HA!
当当当！
KNOCK, KNOCK.
Who's there? Fred.
Fred who? Are you a-fred of ghosts?
谁在敲门？弗雷德。
哪个弗雷德？你怕鬼吗？
*Fred（弗雷德），a-fred谐音afraid（害怕）。

ENCYCLOPEDIA OF ANIMALS 动物小百科

河马是陆地上第三大哺乳动物，仅次于大象和白犀牛。

The hippopotamus is the third largest land mammal after the elephant and white rhinoceros.

Q What kind of dog has no tail?

什么狗没有尾巴？

A A hot dog.

热狗。

当当当！
KNOCK, KNOCK.
Who's there? Thistle.
Thistle who? Thistle be the last time I come here.
谁在敲门？西斯尔。
哪个西斯尔？这将会是我最后一次来这里。
*Thistle（西斯尔）谐音this'll（这将会）。
HA! HA! HA! HA! HA! HA! HA! HA! HA! HA! HA! HA! HA! HA! HA! HA!

海豹是哺乳动物，除了产崽和照顾幼崽时会到陆地上，其余时间都在海里度过。

Seals are mammals that breed and look after their babies on land, but they spend the rest of their time in the ocean.

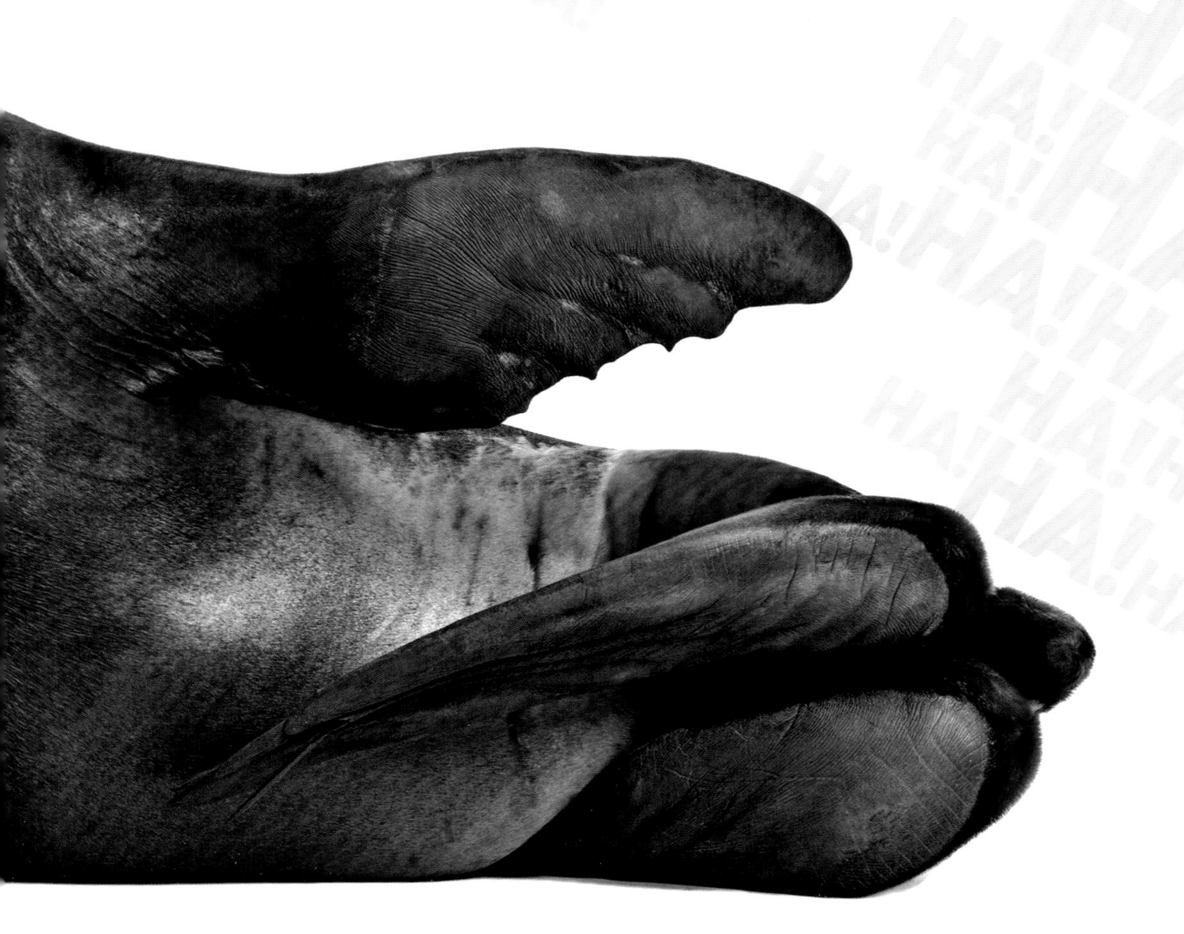

ENCYCLOPEDIA OF ANIMALS 动物小百科

一些变色龙在遇到危险时能够通过改变身体的颜色来伪装自己，或是改变身体的颜色以适应温度变化。

Some chameleons can change their color to camouflage themselves when threatened, or as a reaction to temperature change.

TONGUE TWISTER 英语绕口令

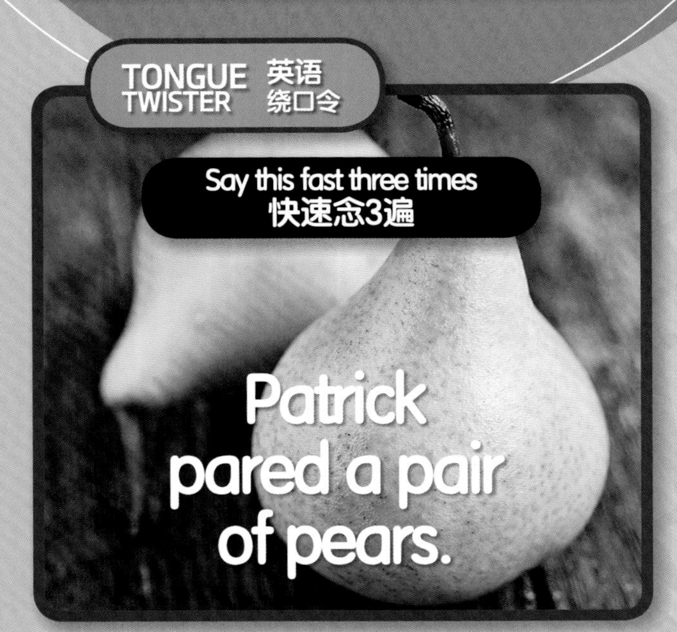

HA! HA! HA! HA! HA! HA! HA! HA! HA! HA!
当当当！
KNOCK, KNOCK.
Who's there? Radio. Radio who? Radio not, here I come!
谁在敲门？雷迪奥。哪个雷迪奥？准备好了吗？我来啦！
*Radio（雷迪奥），radio not谐音ready or not（准备好了吗）。

HA! HA! HA!
当当当！
KNOCK,KNOCK.
Who's there? Gorges.
Gorges who? You look
absolutely gorges!
谁在敲门？戈杰斯。
哪个戈杰斯？你看起
来非常漂亮！
* Gorges（戈杰斯）
谐音gorgeous（极其漂亮的）。

ENCYCLOPEDIA OF ANIMALS 动物小百科

哈士奇的眼睛有几种颜色：蓝色、褐色、绿色或者金色。有些哈士奇一只眼睛是蓝色的，一只眼睛是褐色的。

Huskies can have a range of eye colors: blue, brown, green, or gold. Some have one blue eye and one brown eye.

TONGUE TWISTER 英语绕口令

Say this fast three times
快速念3遍

Furry felines fight fat ferrets.

A!HA!HA!HA!HA!HA!HA!HA!HA!HA!HA!

鹦鹉善于模仿人类语言，这使它们成为珍贵的宠物。有些鹦鹉能活很长时间——长达80年！

Their ability to imitate human speech has made parrots prized pets. Some also live for a long time — up to 80 years!

当当当！
KNOCK,KNOCK.
Who's there? Myth.
Myth who? I myth you too!
谁在敲门？米斯。
哪个米斯？我也想你！
*Myth（米斯）谐音miss（思念）。

狞猫是一名杰出的杂技演员。它可以高高跃起，抓住一只正在飞翔的鸟！

The caracal is an excellent acrobat. It can leap high into the air and catch a bird in flight!